Josiah Coates

Concise Pre–Algebra

To Noah

Contents

Preface

A typical pre-algebra textbook can be over 300 pages long, full of complex theorems and postulates. Working as an engineer for a major international shipping company, I have spent my entire career explaining complex subjects to senior (and highly paid) executives. If I used a "math textbook" approach to communicating with executives, I would have lost my job a long time ago.

Executives, it is believed, are very busy – and they need quick and concise explanations to make quick and well-informed decisions. But are middle and high school students any different? Do they have endless time to read 300 pages of postulates and theorems that have no application to modern life?

To be fair, pre-algebra – along with algebra, geometry and calculus – does have some application to the modern workforce. But it is the general concepts of these subjects that are applicable. The onerous theorems, postulates and definitions – lengthy explanations using incomprehensible terminology are, in my opinion, useless. This approach to math should only be taught to students (college students) that have decided to become mathematicians. And there are quite a small number of mathematicians in the world.

When I prepare papers and presentations on the most complex engineering topics, topics which are often cutting edge and cannot be easily researched on the internet, I continually revise and revise until the presentation is short, concise and understandable by almost anyone. It would seem only logical that a math textbook should be prepared the same way.

The subject matter in this book covers an entire pre-algebra course and can be mastered after about 20 hours of study. It should take an additional 10 hours to master the practice problems at the end of each chapter. The last two chapters are the most difficult, and comprise the core of pre-algebra. I suggest reading these chapters twice.

For any questions or comments, I can be reached at info@concisetextbooks.com. I will respond to any requests for clarification, as I am interested in revising the text in any area where it may be hard to understand.

Sincerely,

Josiah Coates

Chapter 1 Factors

What are the factors of 4?

We know the following:

$$1 \times 4 = 4$$

Therefore, 1 and 4 are factors of 4.

But we also know the following:

$$2 \times 2 = 4$$

Therefore, 2 is also a factor of 4. The factors of 4 are therefore 1, 2 and 4.

What are the factors of 12?

$$1 \times 12 = 12$$

$$2 \times 6 = 12$$

$$3 \times 4 = 12$$

The factors of 12 are: 1, 2, 3, 4, 6 and 12.

As we can see, 1 is always a factor of a number. And the number itself is always a factor. Factors can only be whole numbers but cannot be fractions or decimals.

What are the factors of 24?

$$1 \times 24 = 24$$

$2 \times 12 = 24$

$3 \times 8 = 24$

$4 \times 6 = 24$

The factors of 24 are: 1, 2, 3, 4, 6, 8, 12 and 24.

All factors of a number (except the number itself) cannot be greater than ½ the value of the number.

For example, ½ of 100 is 50. Therefore all factors of 100 (other than 100) cannot be greater than 50.

Another example, all factors of 30 (other than 30), cannot be greater than 15.

The numbers 12 and 24 have a different set of factors. But some of the factors are the same: 1, 2, 3, 4 and 6.

12: 1, 2, 3, 4, 6

24: 1, 2, 3, 4, 6, 8, 12

Therefore, we can say the common factors between 12 and 24 are 1, 2, 3, 4 and 6. Common factors are factors which are shared between two numbers.

What are the common factors of 15 and 30?

15: 1, 3, 5, 15

30: 1, 2, 3, 5, 6, 10, 15, 30

Therefore, the common factors of 15 and 30 are 1, 3, 5 and 15.

The greatest common factor is the largest number that is a factor of two numbers. For example the greatest common factor of 12 and 24 is 6.

What is the greatest common factor of 14 and 35?

14: 1, 2, 7

35: 1, 5, 7, 35

Therefore, the greatest common factor of 14 and 35 is 7.

What is the greatest common factor of 9 and 27?

9: 1, 3, 9

27: 1, 3, 9, 27

Therefore, the greatest common factor of 9 and 27 is 9.

Multiples are similar to factors, but they are the opposite. A multiple goes on forever. Below are the first five multiples of 10:

10, 20, 30, 40, 50

Below are the first five multiples of 5:

5, 10, 15, 20, 25

There are some common multiples between 10 and 5, such as 10 and 20.

What are some common multiples of 6 and 9?

9: 9, 18, 27, 36, 45, 54

6: 6, 12, 18, 24, 30, 36, 42

Therefore, two common multiples of 6 and 9 are 18, and 36.

What are some common multiples of 12 and 18?

12: 12, 24, 36, 48, 60, 72

18: 18, 36, 54, 72, 90

Two common multiples of 12 and 18 are 36 and 72. We are often interested in knowing the least common multiple of two numbers. The least common multiple of 12 and 18 is 36.

What is the least common multiple of 4 and 6?

4: 4, 8, 12, 16, 20

6: 6, 12, 18, 24, 30

As will be learned later in this book, we need to be able to find the greatest common factor and least common multiples of sets of numbers to be able to solve algebra problems.

Practice Problems

1. What are the factors of 6?

Answer: 1, 2, 3, 6

2. What are the factors of 50?

Answer: 1, 2, 5, 10, 25, 50

3. What are the factors of 96?

Answer: 1, 2, 3, 4, 6, 8, 12, 16, 24, 32, 48, 96

4. What are factors of 12?

Answer: 1, 2, 3, 4, 6, 12

5. What are the first five multiples of 2?

Answer: 2, 4, 6, 8, 10

6. What are the first five multiples of 3?

Answer: 3, 6, 9, 12, 15

7. What are the first five multiples of 10?

Answer: 10, 20, 30, 40, 50

8. What are the first five multiples of 9?

Answer: 9, 18, 27, 36, 45

9. What are the first five multiples of 13?

Answer: 13, 26, 39, 52, 65

10. What is the greatest common factor of 6 and 12?

6: 1, 2, 3, 6

12: 1, 2, 3, 4, 6, 12

Answer: 6

11. What is the greatest common factor of 52 and 13?

52: 1, 2, 4, 13, 26, 52

13: 1, 13

Answer: 13

12. What is the greatest common factor of 81 and 72?

81: 1, 3, 9, 27, 81

72: 1, 2, 3, 6, 8, 9, 12, 24, 36, 72

Answer: 9

13. What is the least common multiple of 9 and 6?

9: 9, 18, 27, 36, 45, 54

6: 6, 12, 18, 24, 30

Answer: 18

14. What is the least common multiple of 12 and 15?

12: 12, 24, 36, 48, 60

15: 15, 30, 45, 60, 75

Answer: 60

15. What is the least common multiple of 10 and 25?

10: 10, 20 30, 40, 50, 60 70, 80, 90, 100

25: 25, 50, 75, 100, 125

Answer: 50

Chapter 2 Equations

The expression below is an equation:

$$1 + 2 = 3$$

This expression is also an equation:

$$1 + 2 = 1 + 2$$

The two equations above both express the same thing, but the second equation is more complicated than the first. Below is a slightly more complicated equation:

$$1 + 2 = 1 + 1 + 1$$

And below is an even more complicated equation:

$$3 \times 1 = 1 + 1 + 1$$

The above two equations have expressed the same thing, 3=3. The below equation also expresses 3=3:

$$1 + 1 \times 2 = 1 + 1 + 1$$

How do we know the left side (1+1x2) of the above equation = 3? It appears that we could solve the left side (1+1x2) of the equation two different ways.

First, we could add together the first two terms (1+1) to get 2. This would give us:

$$2 \times 2 = 1 + 1 + 1$$

But this could not be correct, because 2x2 does not equal 1+1+1. Another way to solve the equation is to first multiply 1x2 to get:

$$1 + 2 = 1 + 1 + 1$$

This is correct.

Why are there two different ways to answer the same equation? It turns out that there are special rules for solving equations will addition and multiplication. Under these rules the multiplication must always be performed first. Below is another example:

$$3 \times 2 + 3 = ?$$

Since we now know that multiplication is performed before addition, we can solve the above equation by starting with 3x2=6, which gives us:

$$6 + 3 = 9$$

Below is another example:

$$4 \times 2 + 3 = ?$$

4x2=8, therefore:

$$8 + 3 = 11$$

Below is another example:

$$6 \times 3 + 7 = ?$$

6x3=18, therefore:

$$18 + 7 = 25$$

Below is another more complicated example:

$$2 \times 3 \times 3 = ?$$

Here, it's not clear what we do first. If there are multiple multiplication terms, then we solve each term moving from left to right:

First we solve 2x3 and get:

$$2 \times 3 = 6$$

Then we solve 3x3 and get:

$$2 \times 9 = 18$$

Below is another example:

$$2 \times 3 \times 3 + 2 = ?$$

Here, we must solve all the multiplication first, and then we solve the addition. 2x3=6, therefore:

$$6 \times 3 + 2 = ?$$

6x3=18, therefore:

$$18 + 2 = 20$$

Below is another example:

$$3 \times 2 \times 4 + 4 = ?$$

We first start with the multiplication terms. 3x2=6, therefore:

$$6 \times 4 + 4 = ?$$

Now for the next multiplication term, 6x4=24, therefore:

$$24 + 4 = 28$$

Below is another example:

$$3 \times 3 \times 4 + 2 = ?$$

First we solve the multiplication moving from left to right and then we solve the addition. 3x3=9, therefore:

$$9 \times 4 + 2 = ?$$

9x4=36, therefore:

$$36 + 2 = 38$$

What about problems with division and multiplication? Division is treated the same as multiplication, except each term must be answered in order going from left to right:

$$3 \times 4 \div 2 = ?$$

Since the multiplication comes first going from left to right, we must solve the multiplication term first. Since 3x4=12, we have:

$$12 \div 2 = ?$$

Since 12÷2=6, we have:

$$12 \div 2 = 6$$

Below is another example including multiplication and division:

$$6 \div 3 \times 2 = ?$$

Division is the first term we have moving from left to right. Since 6÷3=2, we have:

$$2 \times 2 = ?$$

Since 2x2=4, we have:

$$2 \times 2 = 4$$

Subtraction is also treated the same as addition. Except the terms should also be answered going from left to right – this is not required but it is a good practice to follow. Below is an example:

$$1 + 3 - 2 = ?$$

Solving from left to right, we have 1+3=4, therefore:

$$4 - 2 = 2$$

Below is another example:

$$7 + 3 - 2 + 8 = ?$$

Solving from left to right, we have 7+3=10, therefore:

$$10 - 2 + 8 = ?$$

Moving along to the next term going from left to right, we have 10-2=8, therefore:

$$8 + 8 = 16$$

Below is another example:

$$12 - 5 + 7 - 5 = ?$$

Solving from left to right, we have 12-5=7, therefore:

$$7 + 7 - 5 = ?$$

Moving along to the next term going from left to right, we have 7+7=14, therefore:

$$14 - 5 = 9$$

Below is another example, this time combining division, multiplication and addition:

$$6 \div 3 + 2 \times 4 = ?$$

First we solve the division terms moving from left to right. We have 6÷3=2, therefore:

$$2 + 2 \times 4 = ?$$

Now we solve the multiplication term. 2x4=8, therefore:

$$2 + 8 = 10$$

Below is another example:

$$4 \div 2 \times 3 + 2 \div 2 = ?$$

First we start with multiplication or division moving from left to right. 4÷2=2, therefore:

$$2 \times 3 + 2 \div 2 = ?$$

Moving along to the next multiplication or division term from left to right, we have 2x3=6, therefore:

$$6 + 2 \div 2 = ?$$

Now we have to skip the addition term and move to the division term. 2÷2=1, therefore:

$$6 + 1 = 7$$

Below is a problem combining division, multiplication, addition and subtraction:

$$2 \times 6 \div 4 + 3 - 8 \div 4 = ?$$

The first multiplication term moving from left to right gives us 2x6=12, therefore:

$$12 \div 4 + 3 - 8 \div 4$$

The next term is 12÷4=3, therefore:

$$3 + 3 - 8 \div 4$$

The next term is 8÷4=2, therefore:

$$3 + 3 - 2$$

The next term is 3+3=6, therefore:

$$6 - 2 = 4$$

Below is another problem combining division, multiplication, addition and subtraction:

$$12 \div 6 \times 2 + 4 - 8 \div 4 - 1 = ?$$

The first multiplication term moving from left to right gives us 12÷6=2, therefore:

$$2 \times 2 + 4 - 8 \div 4 - 1$$

The next term is 2x2=4, therefore:

$$4 + 4 - 8 \div 4 - 1$$

The next term is 8÷4=2, therefore:

$$4 + 4 - 2 - 1$$

The next term is 4+4=8, therefore:

$$8 - 2 - 1$$

The next term is 8-2=6, therefore:

$$6 - 1 = 5$$

There are two more types of terms which must be considered in the order of operations, these are parenthesis and exponents (powers). Parentheses must always be solved first, and exponents (powers) must always be solved second after parenthesis.

Below is an example of an operation with parenthesis:

$$(1 + 3) + 2 \times 2 = ?$$

Here, even though it is an addition term, the (1+3) must come first, therefore:

$$4 + 2 \times 2$$

The next term in the order of operations is 2x2=4, therefore:

$$4 + 4 = 8$$

Below is another example using parenthesis in the order of operations:

$$2 + 6 \times 3 - (8 + 8) \div 4 - 4 = ?$$

The first term we must solve is the parenthesis. (8+8)=16, therefore:

$$2 + 6 \times 3 - 16 \div 4 - 4$$

The next term is 6x3=18, therefore:

$$2 + 18 - 16 \div 4 - 4$$

The next term is 16÷4=4, therefore:

$$2 + 18 - 4 - 4$$

The next term is 2+18=20, therefore:

$$20 - 4 - 4$$

The next term is 20-4=16, therefore:

$$16 - 4 = 12$$

The last type of operation we must consider in the order of operations is exponents, which are also referred to as powers. Exponent operations must be performed after parenthesis. Below is an example:

$$2^3 + 3 \times (1 + 2) = ?$$

The parenthesis must come first. (1+2)=3, therefore:

$$2^3 + 3 \times 3$$

Next is the power (exponents), not the multiplication. 2^3=2x2x2=8, therefore:

$$8 + 3 \times 3$$

The next term is 3x3=9, therefore:

$$8 + 9 = 17$$

Below is another example:

$$3^2 + (16 - 2) \div 7 - 4$$

The first term we must solve is the parenthesis. (16-2)=14, therefore:

$$3^2 + 14 \div 7 - 4$$

The next term is 3^2=9, therefore:

$$9 + 14 \div 7 - 4$$

The next term is 14÷7=2, therefore:

$$9 + 2 - 4$$

The next term is 9+2=11, therefore:

$$11 - 4 = 7$$

If there multiple parenthesis or exponents (powers), then they must be solved in order from left to right. Below is an example:

$$3^2 + (14 - 2) \div (7 - 4) + 2^2$$

The first term we must solve is the first parenthesis from left to right. (14-2)=12, therefore:

$$3^2 + 12 \div (7 - 4) + 2^2$$

The next term is (7-4)=3, therefore:

$$3^2 + 12 \div 3 + 2^2$$

The next term is 3^2=9, therefore:

$$9 + 12 \div 3 + 2^2$$

The next term is 2^2=4, therefore:

$$9 + 12 \div 3 + 4$$

The next term is 12÷3=4, therefore:

$$9 + 4 + 4$$

The next term is 9+4=13, therefore:

$$13 + 4 = 17$$

Therefore the order of operations for solving equations is parenthesis-powers-multiplication-division-addition-subtraction. We can remember this order of operations by using the phrase, Pretty Please My Dear Aunt Sally.

Practice Problems

1. Solve the equation below:

$25 \div 5 + 5 \times 2 - 4 = ?$

Answer:

$5 + 5 \times 2 - 4$

$5 + 10 - 4$

$15 - 4 = 11$

2. Solve the equation below:

$2 - 1 + 11 - 14 = ?$

Answer:

$1 + 11 - 14$

$12 - 14 =- 2$

3. Solve the equation below:

$21 + 11 \times 3 - 6 \times 4 = ?$

Answer:

$21 + 33 - 6 \times 4$

$21 + 33 - 24$

$54 - 24 = 30$

4. Solve the equation below:

$$5 \times 3 + 16 \times 7 - 14 \div 2 + 41 + 3 \div 1 - 9 = ?$$

Answer:

$$15 + 16 \times 7 - 14 \div 2 + 41 + 3 \div 1 - 9$$

$$15 + 112 - 14 \div 2 + 41 + 3 \div 1 - 9$$

$$15 + 112 - 7 + 41 + 3 \div 1 - 9$$

$$15 + 112 - 7 + 41 + 3 - 9$$

$$127 - 7 + 41 + 3 - 9$$

$$120 + 41 + 3 - 9$$

$$161 + 3 - 9$$

$$164 - 9$$

$$155$$

5. Solve the equation below:

$$2^3 \times 3^2 + 4^1 \times 1^4 - 12^0 \div (2 - 1) = ?$$

Answer:

HINT: it looks difficult, but just take it one step at a time.

$$2^3 \times 3^2 + 4^1 \times 1^4 - 12^0 \div 1$$

$$8 \times 3^2 + 4^1 \times 1^4 - 12^0 \div 1$$

$$8 \times 9 + 4^1 \times 1^4 - 12^0 \div 1$$

$8 \times 9 + 4 \times 1^4 - 12^0 \div 1$

$8 \times 9 + 4 \times 1 - 12^0 \div 1$

$8 \times 9 + 4 \times 1 - 1 \div 1$

$72 + 4 \times 1 - 1 \div 1$

$72 + 4 - 1 \div 1$

$72 + 4 - 1$

$76 - 1 = 75$

Chapter 3 Fractions

In order to solve algebraic equations, we must understand how to add, subtract, multiply and divide fractions.

The first step to solving fractions is to reduce the fraction to its lowest terms. A fraction can be reduced to its lowest terms if both the numerator and denominator can be divided evenly by the same number. Below is an example.

Is the fraction below in lowest terms?

$$\frac{4}{8}$$

Answer: No, because both 4 and 8 can be divided by 4 or 2. If we divide both the numerator and denominator by 4, we have:

$$\frac{4 \div 4}{8 \div 4} = \frac{1}{2}$$

Therefore we know that 4/8 = 1/2. We also know that 1/2 is the lowest term form of 4/8.

In order to confirm if a fraction has been reduced to its lowest terms, we must:

1. Find the greatest common factor of the numerator and denominator.
2. Divide the numerator denominator by the greatest common factor.

Reduce the below fraction to its lowest terms:

$$\frac{8}{12}$$

What are factors of 8? 1, 2, 4, 8

What are factors 12? 1, 2, 3, 4, 6

What is the greatest common factor of 8 and 12? 4

Therefore we can divide both the numerator and denominator by 4 to reduce the fraction to its lowest terms:

$$\frac{8 \div 4}{12 \div 4} = \frac{2}{3}$$

Below is another example, reduce the below fraction to its lowest terms:

$$\frac{21}{35}$$

Factors of 21: 1, 3, 7, 21

Factors of 35: 1, 5, 7, 35

Therefore divide both sides by 7:

$$\frac{21 \div 7}{35 \div 7} = \frac{3}{5}$$

Below is one more example, reduce the below fraction to its lowest terms:

$$\frac{15}{25}$$

Factors of 15: 1, 5, 15

Factors of 25: 1, 5, 25

Therefore divide both sides by 5:

$$\frac{15 \div 5}{25 \div 5} = \frac{3}{5}$$

In order to multiply two fractions, we multiply the numerator and denominator separately. See an example below:

$$\frac{2}{7} \times \frac{4}{3} = ?$$

Next, we multiply each term across the numerator and denominator:

$$\frac{2 \times 4}{7 \times 3} = \frac{8}{21}$$

Below is another example:

$$\frac{5}{3} \times \frac{1}{3} = ?$$

Multiply each term across the numerator and denominator:

$$\frac{5 \times 1}{3 \times 3} = \frac{5}{9}$$

Below is another example:

$$\frac{2}{3} \times \frac{3}{4} = ?$$

Multiply each term across the numerator and denominator:

$$\frac{2 \times 3}{3 \times 4} = \frac{6}{12}$$

The answer to the above multiplication problem is 6/12. However, we usually reduce the fraction to lowest terms in the answer.

Factors of 6: 1, 2, 3, 6

Factors of 12: 1, 2, 3, 4, 6, 12

Therefore we can divide both terms by 6, the greatest common factor:

$$\frac{6 \div 6}{12 \div 6} = \frac{1}{2}$$

Dividing fractions is similar to multiplication, except we have to inverse one of the terms. For example:

$$\frac{1}{2} \div \frac{2}{3} = ?$$

Here, we have to inverse one of the terms, and then multiply. It doesn't matter which term is inversed. For this example, we will inverse 2/3:

$$\frac{1}{2} \times \frac{3}{2} = \frac{1 \times 3}{2 \times 2} = \frac{3}{4}$$

Below is another example:

$$\frac{4}{7} \div \frac{3}{4} = ?$$

First we inverse one of the terms, and then we multiply:

$$\frac{4}{7} \times \frac{4}{3} = \frac{4 \times 4}{7 \times 3} = \frac{16}{21}$$

Below is another example:

$$\frac{3}{6} \div \frac{3}{5} = ?$$

First we inverse one of the terms, and then we multiply:

$$\frac{3}{6} \times \frac{5}{3} = \frac{3 \times 5}{6 \times 3} = \frac{15}{18}$$

The answer to the above multiplication problem is 15/18, but we must reduce the fraction to lowest terms. We can reduce 15/18 to lower terms, by finding the factors of each number:

Factors of 15: 1, 3, 5, 15

Factors of 18: 1, 2, 3, 6, 9, 18

Therefore we can divide the terms by 3, the greatest common factor:

$$\frac{15 \div 3}{18 \div 3} = \frac{5}{6}$$

Therefore, the final answer is 5/6.

Add the two fractions below:

$$\frac{1}{4} + \frac{2}{4} = ?$$

Here we can combine the numerators. Since we know 1+2=3, we have:

$$\frac{1+2}{4} = \frac{3}{4}$$

Below is another example:

$$\frac{2}{5} + \frac{2}{5} = ?$$

Here we can combine the numerators. Since we know 2+2=4, we have:

$$\frac{2+2}{5} = \frac{4}{5}$$

Below is another example:

$$\frac{4}{5} + \frac{2}{5} = ?$$

Here we can combine the numerators. Since we know 4+2=6, we have:

$$\frac{4+2}{5} = \frac{6}{5} = 1\frac{1}{5}$$

In the preceding problems, we have been adding fractions with the same denominators. In the below example, we show how to add fractions with different denominators:

$$\frac{2}{3} + \frac{3}{4}$$

First we cross multiply:

$$3 \times 3 = 9$$

$$2 \times 4 = 8$$

Then we add the two answers together:

$$9 + 8 = 17$$

This is the numerator of the answer. For the denominator, we multiply the two denominators to get:

$$3 \times 4 = 12$$

Now we know the numerator is 17 and the denominator is 12, therefore the answer is:

$$\frac{17}{12}$$

Below is another example:

$$\frac{1}{2} + \frac{2}{3}$$

First we cross multiply:

$$2 \times 2 = 4$$

$$1 \times 3 = 3$$

Then we add the two together:

$$3 + 4 = 7$$

Therefore 7 is the numerator, now we multiply the two denominators to get the denominator:

$$2 \times 3 = 6$$

Therefore the answer is:

$$\frac{7}{6}$$

Below is another example:

$$\frac{2}{6} + \frac{1}{4}$$

First we cross multiply:

$$6 \times 1 = 6$$

$$2 \times 4 = 8$$

Then we add the two together:

$$6 + 8 = 14$$

Therefore 14 the numerator, now we multiply the two denominators to get the new denominator:

$$6 \times 4 = 24$$

This gives us:

$$\frac{14}{24}$$

We can reduce this fraction to its lowest terms:

Factors of 14: 1, 2, 7, 14

Factors of 24: 1, 2, 3, 4, 6, 8, 12, 24

The greatest common factor is 2, therefore we can divide the fraction by two:

$$\frac{14 \div 2}{24 \div 2} = \frac{7}{12}$$

Subtract the two fractions below:

$$\frac{2}{3} - \frac{1}{3} = ?$$

Since they have the same denominator, we can combine the numerators into the same fraction:

$$\frac{2 - 1}{3} = \frac{1}{3}$$

Below is another example:

$$\frac{3}{5} - \frac{1}{5} = ?$$

Since they have the same denominator, we can combine the numerators into the same fraction:

$$\frac{3-1}{5} = \frac{2}{5}$$

Below is another example:

$$\frac{2}{3} - \frac{2}{5} = ?$$

When the fractions have different denominators, we cannot follow the same rules as addition. We start by finding the least common multiple of each denominator:

3: 3, 6, 9, 12, 15

5: 5, 10, 15, 20, 25

The least common multiple is 15. This will be the new denominator for both of the "new" fractions. Below is how we find the "new" numerator for 2/3:

$$\frac{2}{3} = \frac{?}{15}$$

First we divide the "new" denominator (15) by the old denominator (3):

$$15 \div 3 = 5$$

Now we cross multiply this answer by the "old" numerator (2):

$$2 \times 5 = 10$$

The new numerator is 10, therefore:

$$\frac{2}{3} = \frac{10}{15}$$

Now we must convert 2/5 to the new fraction:

$$\frac{2}{5} = \frac{?}{15}$$

First divide the "new" denominator (15) by the "old" denominator (5):

$$15 \div 5 = 3$$

Now we cross multiply this answer by the "old" numerator (2):

$$2 \times 3 = 6$$

The new numerator is 6, therefore:

$$\frac{2}{5} = \frac{6}{15}$$

Now we return to the original equation:

$$\frac{2}{3} - \frac{2}{5} = ?$$

We can replace each fraction with the new fractions:

$$\frac{10}{15} - \frac{6}{15} = \frac{4}{15}$$

4/15 is the correct answer. Below is another example:

$$\frac{3}{5} - \frac{3}{7} = ?$$

The first step is to find the least common multiple of each denominator:

7: 7, 14, 21, 28, 35, 42

5: 5, 10, 15, 20, 25, 30, 35

The least common multiple is 35. Therefore, the new denominator is 35. Now we must convert 3/5 to the new fraction:

$$\frac{3}{5} = \frac{?}{35}$$

First divide the "new" denominator (35) by the "old" denominator (5):

$$35 \div 5 = 7$$

Now we cross multiply this answer by the "old" numerator (3):

$$3 \times 7 = 21$$

The new numerator is 21, therefore:

$$\frac{3}{5} = \frac{21}{35}$$

Now to calculate the new numerator for 3/7:

$$\frac{3}{7} = \frac{?}{35}$$

First divide the "new" denominator (35) by the "old" denominator (7):

$$35 \div 7 = 5$$

Now we cross multiply this answer by the "old" numerator (3):

$$3 \times 5 = 15$$

The new numerator is 15, therefore:

$$\frac{3}{7} = \frac{15}{35}$$

Now we return to the original equation:

$$\frac{3}{5} - \frac{3}{7} = ?$$

We can replace each fraction with the new fractions:

$$\frac{21}{35} - \frac{15}{35} = \frac{6}{35}$$

6/35 is the correct answer. Below is another example:

$$\frac{7}{9} - \frac{1}{2} = ?$$

The first step is to find the least common multiple of each denominator:

9: 9, 18, 27, 36

2: 2, 4, 6, 8, 10, 12, 14, 16, 18

The least common multiple is 18. Therefore, the new denominator is 18. Now we must convert 7/9 to the new fraction:

$$\frac{7}{9} = \frac{?}{18}$$

First divide the "new" denominator (18) by the "old" denominator (9):

$$18 \div 9 = 2$$

Now we cross multiply this answer by the "old" numerator (7):

$$2 \times 7 = 14$$

The new numerator is 14, therefore:

$$\frac{7}{9} = \frac{14}{18}$$

Now to calculate the new numerator for 1/2:

$$\frac{1}{2} = \frac{?}{18}$$

First divide the "new" denominator (18) by the "old" denominator (2):

$$18 \div 2 = 9$$

Now we cross multiply this answer by the "old" numerator (1):

$$1 \times 9 = 9$$

The new numerator is 9, therefore:

$$\frac{1}{2} = \frac{9}{18}$$

Now we return to the original equation:

$$\frac{7}{9} - \frac{1}{2} = ?$$

We can replace each fraction with the new fractions:

$$\frac{14}{18} - \frac{9}{18} = \frac{5}{18}$$

5/18 is the correct answer.

We will conclude this chapter after a brief discussion on mixed numbers. Mixed numbers are fractions that include both whole numbers and fractions, such as:

$$1\frac{1}{2}$$

We know that the following is true:

$$\frac{3}{2} = 1\frac{1}{2}$$

On some occasions problems will be presented with fractions as mixed numbers such as 1 ½ instead of 3/2. We generally recommend to convert mixed numbers to fractions before performing any operations such as adding, multiplying or dividing.

The use of mixed numbers instead of fractions is become exceedingly rare in the professional world, and the time spent studying specific rules for mixed numbers, versus converting to fractions, should be accordingly minimal.

Practice Problems

1. Is the below fraction reduced to its lowest terms?

$$\frac{2}{4}$$

Answer: No, both numbers are divisible by 2:

$$\frac{2 \div 2}{4 \div 2} = \frac{1}{2}$$

2. Is the below fraction reduced to its lowest terms?

$$\frac{3}{9}$$

Answer: No, both numbers are divisible by 3:

$$\frac{3 \div 3}{9 \div 3} = \frac{1}{3}$$

3. Is the below fraction reduced to its lowest terms?

$$\frac{4}{9}$$

Answer: Yes, while 4 is divisible by 2, and 9 is divisible by 3, neither number is divisible by the same number (other than 1), therefore the fraction is already reduced to its lowest terms.

4. Is the below fraction reduced to its lowest terms?

$$\frac{44}{70}$$

Answer: if the numerator and denominator are both even, then we know that the fraction is not reduced to its lowest terms, because each term can be divided by 2:

$$\frac{44 \div 2}{70 \div 2} = \frac{22}{35}$$

5. Is the below fraction reduced to its lowest terms?

$$\frac{45}{108}$$

Answer: No, both the numerator and denominator are divisible by 9:

$$\frac{45 \div 9}{108 \div 9} = \frac{5}{12}$$

6. Solve the problem below:

$$\frac{2}{7} \times \frac{4}{3} = ?$$

Answer:

$$\frac{2}{7} \times \frac{4}{3} = \frac{2 \times 4}{7 \times 3} = \frac{8}{21}$$

7. Solve the problem below:

$$\frac{4}{3} \times \frac{6}{8} = ?$$

Answer:

$$\frac{4}{3} \times \frac{6}{8} = \frac{4 \times 6}{3 \times 8} = \frac{24}{24} = 1$$

8. Solve the problem below:

$$\frac{8}{3} \times \frac{6}{8} = ?$$

Answer:

$$\frac{8}{3} \times \frac{6}{8} = \frac{8 \times 6}{3 \times 8} = \frac{48}{24} = 2$$

9. Solve the problem below:

$$\frac{2}{11} \times \frac{17}{3} = ?$$

Answer:

$$\frac{2}{11} \times \frac{17}{3} = \frac{2 \times 17}{11 \times 3} = \frac{34}{33}$$

10. Solve the problem below:

$$\frac{1}{2} \div \frac{2}{3} = ?$$

Answer:

$$\frac{1}{2} \times \frac{3}{2} = \frac{1 \times 3}{2 \times 2} = \frac{3}{4}$$

11. Solve the problem below:

$$\frac{3}{4} \div \frac{5}{9} = ?$$

Answer:

$$\frac{3}{4} \div \frac{5}{9} = \frac{3 \times 9}{4 \times 5} = \frac{27}{20}$$

12. Solve the problem below:

$$\frac{7}{17} \div \frac{3}{8} = ?$$

Answer:

$$\frac{7}{17} \div \frac{3}{8} = \frac{7 \times 8}{17 \times 3} = \frac{56}{51}$$

13. Solve the problem below:

$$\frac{1}{2} + \frac{2}{3}$$

First we cross multiply:

$2 \times 2 = 4$

$1 \times 3 = 3$

Then we add the two together:

$3 + 4 = 7$

This is the numerator, now we multiply the two denominators to get the new denominator:

$2 \times 3 = 6$

Therefore the answer is:

$$\frac{7}{6}$$

14. Solve the problem below:

$$\frac{1}{3} + \frac{3}{7}$$

First we cross multiply:

$3 \times 3 = 9$

$1 \times 7 = 7$

Then we add the two together:

$9 + 7 = 16$

This is the numerator, now we multiply the two denominators to get the new denominator:

$$3 \times 7 = 21$$

Therefore the answer is:

$$\frac{16}{21}$$

15. Solve the problem below:

$$\frac{4}{17} + \frac{5}{9}$$

First we cross multiply:

$$5 \times 17 = 85$$

$$4 \times 9 = 36$$

Then we add the two together:

$$85 + 36 = 121$$

This is the new numerator, now we multiply the two denominators to get the new denominator:

$$17 \times 9 = 153$$

Therefore the answer is:

$$\frac{121}{153}$$

16. Solve the problem below:

$$\frac{9}{8} + \frac{2}{3}$$

First we cross multiply:

$$8 \times 2 = 16$$

$$9 \times 3 = 27$$

Then we add the two together:

$$16 + 27 = 43$$

This is the numerator, now we multiply the two denominators to get the new denominator:

$$8 \times 3 = 24$$

Therefore the answer is:

$$\frac{43}{24}$$

17. Solve the problem below:

$$\frac{2}{3} - \frac{1}{3} = ?$$

Since they have the same denominator, we can combine the numerators into the same fraction:

$$\frac{2-1}{3} = \frac{1}{3}$$

18. Solve the problem below:

$$\frac{2}{3} - \frac{3}{7} = ?$$

Since we are subtracting with different denominators, the first step is to find the least common multiple of each denominator:

3: 3, 6, 9, 12, 15, 18, 21

7: 7, 14, 21, 28, 35, 42

Therefore, the new denominator is 21. Now we must convert 2/3 to the new fraction:

$$\frac{2}{3} = \frac{?}{21}$$

First divide the "new" denominator by the "old" denominator:

$$21 \div 3 = 7$$

Now we cross multiply this answer by the "old" numerator:

$$2 \times 7 = 14$$

The new numerator is 14, therefore:

$$\frac{14}{21} = \frac{2}{3}$$

Now to calculate the new numerator for 3/7:

$$\frac{3}{7} = \frac{?}{21}$$

First divide the "new" denominator by the "old" denominator:

$$21 \div 7 = 3$$

Now we cross multiply this answer by the "old" numerator:

$$3 \times 3 = 9$$

The new numerator is 9, therefore:

$$\frac{3}{7} = \frac{9}{21}$$

Now we return to the original equation:

$$\frac{2}{3} - \frac{3}{7} = ?$$

We can replace each fraction with the new fractions:

$$\frac{14}{21} - \frac{9}{21} = \frac{5}{21}$$

19. Solve the problem below:

$$\frac{5}{8} - \frac{7}{12} = ?$$

Least common multiple of each denominator:

8: 8, 16, 24, 32, 40

12: 12, 24, 36, 48, 60

The new denominator is 24:

$$\frac{5}{8} = \frac{?}{24}$$

Now to find the numerator for 5/8, first divide:

$$24 \div 8 = 3$$

Then cross multiply:

$$3 \times 5 = 15$$

The new numerator is 15, therefore:

$$\frac{15}{24} = \frac{5}{8}$$

Now to calculate the new numerator for 7/12:

$$\frac{7}{12} = \frac{?}{24}$$

First divide:

$$24 \div 12 = 2$$

Now we cross multiply this answer by the "old" numerator:

$$7 \times 2 = 14$$

The new numerator is 14, therefore:

$$\frac{7}{12} = \frac{14}{24}$$

Now we return to the original equation:

$$\frac{5}{8} - \frac{7}{12} = ?$$

We can replace each fraction with the new fractions:

$$\frac{15}{24} - \frac{14}{24} = \frac{1}{24}$$

20. Solve the problem below:

$$\frac{5}{3} - \frac{3}{5} = ?$$

Least common multiple of each denominator:

3: 3, 6, 9, 12, 15

5: 5, 10, 15, 20

The new denominator is 15:

$$\frac{5}{3} = \frac{?}{15}$$

Now to find the numerator for 5/3, first divide:

$$15 \div 3 = 5$$

Then cross multiply:

$$5 \times 5 = 25$$

The new numerator is 25, therefore:

$$\frac{25}{15} = \frac{5}{3}$$

Now to calculate the new numerator for 3/5:

$$\frac{3}{5} = \frac{?}{15}$$

First divide:

$$15 \div 5 = 3$$

Now we cross multiply this answer by the "old" numerator:

$$3 \times 3 = 9$$

The new numerator is 9, therefore:

$$\frac{3}{5} = \frac{9}{15}$$

Now we return to the original equation:

$$\frac{5}{3} - \frac{3}{5} = ?$$

We can replace each fraction with the new fractions:

$$\frac{25}{15} - \frac{9}{15} = \frac{16}{15}$$

21. Solve the problem below:

$$\frac{3}{4} - \frac{6}{11} = ?$$

Least common multiple of each denominator:

4: 4, 8, 12, 16, 20, 24, 28, 32, 36, 40, 44

11: 11, 22, 33, 44, 55, 66

The least common multiple is 44. Therefore, the new denominator is 44:

$$\frac{3}{4} = \frac{?}{44}$$

Now to find the numerator for 3/4, first divide:

$$44 \div 4 = 11$$

Then cross multiply:

$$3 \times 11 = 33$$

The new numerator is 33, therefore:

$$\frac{33}{44} = \frac{3}{4}$$

Now to calculate the new numerator for 6/11:

$$\frac{6}{11} = \frac{?}{44}$$

First divide:

$$44 \div 11 = 4$$

Now we cross multiply this answer by the "old" numerator:

$$6 \times 4 = 24$$

The new numerator is 6, therefore:

$$\frac{6}{11} = \frac{24}{44}$$

Now we return to the original equation:

$$\frac{3}{4} - \frac{6}{11} = ?$$

We can replace each fraction with the new fractions:

$$\frac{33}{44} - \frac{24}{44} = \frac{9}{44}$$

Chapter 4 Decimals and Percentage

In this chapter we will review the procedures to add, subtract, multiply and divide fractions.

Below is the first example:

$$10.1 + 9.556 = ?$$

The first step to adding decimal numbers is to align the decimal point:

$$
\begin{array}{r}
10.1 \\
+9.556 \\
\hline
\end{array}
$$

For 10.1, since there is no number after the "1" we assume that the numbers to the right of "1" are all 0. Then we add each number vertically:

$$
\begin{array}{r}
10.100 \\
+9.556 \\
\hline
19.656 \\
\end{array}
$$

Below is another example:

$$
\begin{array}{r}
17.52 \\
+3.476 \\
\hline
20.996 \\
\end{array}
$$

Below is another example:

$20.676 - 15.457 = ?$

We follow the same rule for addition when we are subtracting decimals:

$$\begin{array}{r} 20.676 \\ -15.457 \\ \hline 5.219 \end{array}$$

Below is another example:

$$\begin{array}{r} 89.67 \\ -47.662 \\ \hline 42.008 \end{array}$$

When multiplying decimal points, the first step is to perform the multiplication as if there are no decimal points:

$$\begin{array}{r} 2.11 \\ \times 2.1 \\ \hline 211 \\ +4220 \\ \hline 4431 \end{array}$$

Now we count the number of digits to the right of each decimal point. 2.11 has two digits to the right of the decimal point, and 2.1 has one digit to the right of the decimal point. Therefore there are three digits total to the right of the decimals. Therefore, we take the "final" answer (4431), and put in a decimal point which has three digits to the right:

4.431

Below is another example:

$$
\begin{array}{r}
3.56 \\
\underline{\times 1.4} \\
1424 \\
\underline{+3560} \\
4984
\end{array}
$$

Since there are three digits to the right the decimal point, the final answer must have three digits to the right of the decimal point:

$$4.984$$

Below is another example:

$$
\begin{array}{r}
2.716 \\
\underline{\times 3.454} \\
10864 \\
135800 \\
1086400 \\
\underline{+8148000} \\
9.381064
\end{array}
$$

Below is a division problem with decimals:

$$2.1 \,\overline{\left)8.4\right.}$$

When dividing decimal numbers, we first make the divisor (2.1) a whole number by moving the decimal point to the right. If we move the decimal point one space in the divisor (2.1), then we must also move the decimal point one space in the dividend:

$$21.\overline{\smash{\big)}\,84.}$$

The decimal point for the quotient (answer), must always line up with the decimal point for the dividend (84):

$$21.\overline{\smash{\big)}\,84.}\;\;\overset{\textstyle .}{}$$

84 divided by 21 equals 4.0, therefore:

$$21.\overline{\smash{\big)}\,84.}\;\;\overset{\textstyle 4.}{}$$

Below is another example:

$$2.4\overline{\smash{\big)}\,10.8}$$

Here, we must first make the divisor (2.4) a whole number by moving one decimal point. Therefore, we must also move one decimal point for the dividend (10.8). This gives us:

$$24.\overline{\smash{\big)}\,108.}$$

24 divides into 108 four times. Therefore, we have a 4 in the quotient:

$$\begin{array}{r} 4. \\ 24.\overline{\big)\,108.} \end{array}$$

But 24 does not divide evenly into 108, because there is a remainder. Now we must multiply 4 by 24, and subtract this amount from 108, which gives us:

$$\begin{array}{r} 4. \\ 24.\overline{\big)\,108.} \\ -\,96 \\ \hline 12 \end{array}$$

Now we add a "zero" to 12 which gives us:

$$\begin{array}{r} 4. \\ 24.\overline{\big)\,108.} \\ -\,96 \\ \hline 120 \end{array}$$

Now we divide 120 by 24:

$$\begin{array}{r} 4.5 \\ 24.\overline{\big)\,108.} \\ -\,96 \\ \hline 120 \end{array}$$

4.5 is the final answer. Therefore, 10.8 divided by 2.4 is 4.5.

Below is another example:

$$.005\overline{\smash{\big)}\ 10}$$

First we must make the divisor (.005) a whole number by moving the decimal point three spaces to the right. When we do this, we must also insert three digits to the right of the divisor:

$$5\overline{\smash{\big)}\ 10{,}000}$$

Now we divide 10,000 by 5:

$$5\overline{\smash{\big)}\ 10{,}000}^{\ \ 2{,}000}$$

Convert the below decimal to a fraction:

0.01

Proceed as follows:

$$0.01 = \frac{0.01}{1}$$

Now we add digits to the right of both the numerator and denominator, until both are whole numbers:

$$\frac{0.01}{1} = \frac{0.10}{10} = \frac{1.00}{100}$$

Therefore:

$$0.01 = \frac{1}{100}$$

Convert the below decimal to a fraction:

0.13

Divide the fraction:

$$\frac{.13}{1} = \frac{1.3}{10} = \frac{13}{100}$$

Therefore:

$$.13 = \frac{13}{100}$$

Convert the below decimal to a fraction:

0.271

Divide the fraction:

$$\frac{.271}{1} = \frac{2.71}{10} = \frac{27.1}{100} = \frac{271}{1,000}$$

Therefore:

$$.271 = \frac{271}{1,000}$$

Convert the below decimal to a fraction:

0.25

Divide the fraction:

$$\frac{.25}{1} = \frac{2.5}{10} = \frac{25}{100}$$

Since 25 is a factor of 100, we can divide both sides by 25:

$$\frac{25 \div 25}{100 \div 25} = \frac{1}{4}$$

Therefore:

$$.25 = \frac{1}{4}$$

Convert the below decimal to a fraction:

0.35

Divide the fraction:

$$\frac{.35}{1} = \frac{3.5}{10} = \frac{35}{100}$$

We should determine the factors of 35 and 100 to determine if 35/100 is reduced to its lowest terms.

Factors of 35: 1, 5, 7, 35

Factors of 100: 1, 2, 4, 5, 10, 20, 25, 50, 100

Therefore we can divide both sides by 5:

$$\frac{35 \div 5}{100 \div 5} = \frac{7}{20}$$

Therefore:

$$0.35 = \frac{7}{20}$$

Convert the below fraction to a decimal:

$$\frac{1}{2}$$

To convert a fraction to a decimal, we divide the numerator by the denominator:

$$1 \div 2 = 0.5$$

Therefore:

$$\frac{1}{2} = 0.5$$

Convert the below fraction to a decimal:

$$\frac{3}{5}$$

Divide the numerator by the denominator:

$$5 \overline{)\,3.0\,}^{\,.6}$$

Therefore:

$$\frac{3}{5} = 0.6$$

Convert the below fraction to a decimal:

$$\frac{5}{8}$$

Divide the numerator by the denominator:

$$
\begin{array}{r}
.625 \\
8 \overline{)\,5.0\,} \\
-\ 48 \\
\hline
20 \\
-\ 16 \\
\hline
40
\end{array}
$$

Therefore:

$$\frac{5}{8} = .625$$

Percentages are essentially the same as decimals. For example:

$$100\% = 1.0$$

$50\% = 0.50$

$25\% = 0.25$

Therefore, all the rules of addition, subtraction, multiplication and division that apply to decimals, also apply to percentages. For example:

$50\% + 50\% = 100\%$

Is the same thing as:

$0.5 + 0.5 = 1.0$

If 100% = 1, then we can easily conclude that 300% = 3, and that 1,000% = 10.

To solve a percentage problem, we convert the percentage to decimals and then solve using the previous lessons for solving decimal problems. For example:

$350\% \times 250\% = ?$

First we convert each percentage to decimal:

$3.5 \times 2.5 = ?$

And then we solve:

$$
\begin{array}{r}
3.5 \\
\times 2.5 \\
\hline
175 \\
+700 \\
\hline
8.75
\end{array}
$$

Remember:

$8.75 = 875\%$

Therefore:

$350\% \times 250\% = 875\%$

Practice Problems

1. What is 14.574 + 8.716?

Answer:

$$\begin{array}{r} 14.574 \\ +8.716 \\ \hline 23.290 \end{array}$$

2. What is 72.850 − 58.852:

Answer:

$$\begin{array}{r} 72.850 \\ -58.852 \\ \hline 13.998 \end{array}$$

3. What is 52.473 − 9.378?

Answer:

$$\begin{array}{r} 52.473 \\ -9.378 \\ \hline 43.095 \end{array}$$

4. What is 4.568 x 6.491?

Answer:

$$
\begin{array}{r}
4.568 \\
\underline{x6.491} \\
4568 \\
411120 \\
1827200 \\
\underline{+27408000} \\
29.650888
\end{array}
$$

5. What is 34 ÷ .00056? *(round the answer to 4 decimal points)*

Answer:

$$
.00056\overline{\smash{\big)}\,34}
$$

First move the decimal point:

$$
56\overline{\smash{\big)}\,3400000.0000}
$$

Then solve:

$$
\begin{array}{r}
60714.2855 \\
56\ \overline{)\ 3400000.0000} \\
-\ 336 \\
\hline
400 \\
-\ 392 \\
\hline
80 \\
-\ 56 \\
\hline
240 \\
-\ 224 \\
\hline
160 \\
-\ 112 \\
\hline
480 \\
-\ 448 \\
\hline
320 \\
-\ 280 \\
\hline
400
\end{array}
$$

6. Convert the below decimal to a fraction:

0.42

Divide the fraction:

$$
\frac{.42}{1} = \frac{4.2}{10} = \frac{42}{100}
$$

However, 42/100 is not reduced to its lowest terms, since both numbers can be divided by 2:

$$\frac{42 \div 2}{100 \div 2} = \frac{21}{50}$$

Therefore:

$$0.42 = \frac{21}{50}$$

7. Convert the below decimal to a fraction:

0.25

Divide the fraction:

$$\frac{.25}{1} = \frac{2.5}{10} = \frac{25}{100}$$

However, 25/100 is not reduced to its lowest terms, since both numbers can be divided by 25:

$$\frac{25 \div 25}{100 \div 25} = \frac{1}{4}$$

8. Convert the below fraction to a decimal:

$$\frac{2}{5} = ?$$

Answer:

$$5 \overline{)2.0} \quad .4$$

Therefore:

$$\frac{2}{5} = .4$$

9. Convert the below fraction to a decimal:

$$\frac{3}{4} = ?$$

Answer:

$$
\begin{array}{r}
.75 \\
4 \overline{)3.0} \\
-\ 28 \\
\hline
20 \\
-\ 20 \\
\hline
0
\end{array}
$$

Therefore:

$$\frac{3}{4} = .75$$

10. What is 100% + 50%?

Answer:

$$100\% + 50\% = 150\%$$

11. What is 140% + 370%?

Answer:

$$140\% + 370\% = 510\%$$

12. What is 100% x 50%?

Answer:

$$100\% \times 50\% = 50\%$$

13. What is 300% x 400%?

Answer:

$$300\% \times 400\% = 1200\%$$

Chapter 5 Algebraic Expressions

This is the most difficult chapter of this book. Here, we are no longer learning basic math operations. This chapter is pure algebra, and learning how to solve algebraic expressions. The key to solving algebraic expressions is to remember that while each problem may appear to be long and complicated, it is just a lot of simple math operations put together into one problem.

In Chapter 2 we reviewed equations. The first equation we reviewed was:

$$1 + 2 = 3$$

If we replaced one of the numbers with a letter, we would have created an algebraic expression. For example:

$$1 + x = 3$$

An algebraic equation exists when some of the numbers in an equation are replaced with letters. Then we must determine what value for the letters will allow for the equation to be true. Here, we know that if x=2, the equation above will be true.

In other words, we balance the equation when we determine that x=2. This is the meaning of the word algebra, which is balance. The root of algebra is calculating the values that balance an equation.

There is enormous value in understanding the basics of algebra, not only in the professional world, but in normal day-to-day activities. From financial calculations, to projections, bookkeeping, basic engineering and scientific calculations.

The "letters" in an algebraic equation are generally referred to as variables, and the "numbers" are referred to as constants. The most common variable is "x", although "y" and "z" are also fairly common.

In this chapter we will discuss rules and procedures for manipulating algebraic equations. In the next chapter we will discuss solving (i.e. balancing) algebraic equations.

One of the first steps to solving algebraic equations is to combine like terms. Below is an example:

$$2x + 3y - 6x + 4x^2 + 5y^2 = 4x - 3x^2 + 4y^2 + 2$$

So which terms are "like terms"? There are five different sets of like terms in the equation above. Below is a list of each of the like terms:

$2x, \ -6x, 4x$

$3y$

$4x^2, \ -3x^2$

$5y^2, 4y^2$

2

As you can see, "like terms" are separated by their variables (x or y) and the exponents on the variables. Therefore x and y are not like terms, x and x^2 are not like terms, and y and y^2 are not like terms. And a constant, in this case "2" is not a like term with anything but another constant.

Below is another example:

$$3x + 2x^2 + 3 - x + 3x^2 + 1$$

There are three different sets of like terms in the above equation, what are the like terms?

$$3x, \ -x$$

$$2x^2, 3x^2$$

$$3, 1$$

Below is another example:

$$5y + 8x + 7y + 3 + 3x^2 + y$$

There are four different sets of like terms in the above equation, what are the like terms?

$$5y, 7y, y$$

$$3x^2$$

$$8x$$

$$3$$

After the like terms have been identified, they must be combined. For example, we combine like terms below:

$$3x + 4y - 2x + 2y$$

Like terms:

$$3x, \; -2x$$

$$4y, \; 2y$$

Combine like terms:

$$3x - 2x = x$$

$$4y + 2y = 6y$$

Therefore, the equation once the terms are combined is:

$$x + 6y$$

Combine like terms in the equation below:

$$6x + 3y - 3x + y$$

Answer:

$$3y + y = 4y$$

$$6x - 3x = 3x$$

Therefore:

$$6x + 3y - 3x + y = 4y + 3x$$

Combine like terms in the equation below:

$$4x + 8x^2 - 8x + x^2$$

Answer:

$$4x - 8x = -4x$$

$$8x^2 + x^2 = 9x^2$$

Therefore:

$$4x + 8x^2 - 8x + x^2 = -4x + 9x^2$$

Combine like terms in the equation below:

$$9x^2 - 5y^2 + 3x^2 + 4y^2 + 3$$

Answer:

$$9x^2 + 3x^2 = 12x^2$$

$$-5y^2 + 4y^2 = -y^2$$

Therefore:

$$9x^2 - 5y^2 + 3x^2 + 4y^2 + 3 = 12x^2 - y^2 + 3$$

Now we add another step, we want to solve the equation, given a certain value for x and y.

Solve the below equation, with x=1 and y=2:

$$6x - 3y + 7x + 7y + 5$$

First we combine like terms:

$$6x + 7x = 13x$$

$$-3y + 7y = 4y$$

Therefore:

$$13x + 4y + 5$$

Substitute x=1 and y=2:

$$13(1) + 4(2) + 5$$

$$13 + 8 + 5$$

$$26$$

Below is another example, solve the below equation, with x=2 and y=4:

$$4y - 8x + 4x + y + 8$$

First we combine like terms:

$$-8x + 4x = -4x$$

$$4y + y = 5y$$

Therefore:

$$-4x + 5y + 8$$

Substitute x=2 and y=4:

$$-4(2) + 5(4) + 8$$

$$-8 + 20 + 8$$

$$20$$

Below is another example, solve the below equation, with x=3 and y=5:

$$5y + 6x + 2x + 8y + 2$$

First we combine like terms:

$$6x + 2x = 8x$$

$$5y + 8y = 13y$$

Therefore:

$$8x + 13y + 2$$

Substitute x=3 and y=5:

$$8(3) + 13(5) + 2$$

$$24 + 65 + 2$$

$$91$$

So far we have only been adding and subtracting, below is an example for multiplication:

$$3y \times 2x = \,?$$

Here we can multiply different variables, therefore:

$$3y \times 2x = 3 \cdot 2yx = 6yx$$

Notice the difference here, we can multiply 3y and 2x, but we cannot add or subtract 3y and 2x. This is because we can never add or subtract unlike terms, but we can multiply and divide them.

Below is another example of multiplying terms:

$$4x \times 5y = \,?$$

Here we can multiply different variables, therefore:

$$4x \times 5y = 4 \cdot 5xy = 20xy$$

Below is another example of multiplying terms (when we use parenthesis, it is the same as using a multiplication sign):

$$3y(6x) = \,?$$

Here we can multiply different variables, therefore:

$$3y \times 6x = 3 \cdot 6yx = 18yx$$

Solve the below problem:

$$x \cdot x = \,?$$

Anything times itself, is always the same number squared:

$$x \cdot x = x^{1+1} = x^2$$

Solve the below problem:

$$x^2 \cdot x = \,?$$

Here, we always add the exponents:

$$x^2 \cdot x = x^{2+1} = x^3$$

Below is another example:

$$y^3 \cdot y = \,?$$

Answer:

$$y^3 \cdot y = y^{3+1} = y^4$$

Below is another example:

$$z^2 \cdot z^2 \cdot z^4 = \,?$$

Answer:

$$z^2 \cdot z^2 \cdot z^4 = z^{2+2+4} = z^8$$

The problem below is identical to the problem above, except each of the variables are different:

$$x^2 \cdot y^2 \cdot z^4 = \,?$$

When we multiply different variables, we cannot add the exponents. Therefore:

$$x^2 \cdot y^2 \cdot z^4 = x^2 y^2 z^4$$

Below is another example:

$$x^4 \cdot y^5 \cdot z^2 \cdot x^2 = ?$$

Here, we have two common variables and two different variables, therefore we add exponents for x but not for y and z:

$$x^4 \cdot y^5 \cdot z^2 \cdot x^2 = x^{4+2} y^5 z^2 = x^6 y^5 z^2$$

Below is another example:

$$x^4 \cdot y^5 \cdot z^2 \cdot z^2 = ?$$

Answer:

$$x^4 \cdot y^5 \cdot z^2 \cdot z^2 = x^4 y^5 z^{2+2} = x^4 y^5 z^4$$

Below is another example:

$$x^4 \cdot y^5 \cdot y^2 \cdot z^2 = ?$$

Answer:

$$x^4 \cdot y^5 \cdot y^2 \cdot z^2 = x^4 y^{5+2} z^2 = x^4 y^7 z^2$$

Below is another example:

$$2x^4 \cdot 2z^2 = ?$$

Here, we multiply the constants (2) together:

$$2 \cdot 2 \cdot x^4 \cdot z^2 = 4x^4z^2$$

Below is another example:

$$2x^5 \cdot 3y^3 = ?$$

Answer:

$$2 \cdot 3 \cdot x^5 \cdot y^3 = 6x^5y^3$$

Solve the division problem below:

$$\frac{x}{x} = ?$$

The same number divided by itself is always one. The same is true for variables. Therefore:

$$\frac{x}{x} = 1$$

Below is another example:

$$\frac{2x}{x} = ?$$

The rules for division are the same as multiplication. Therefore, we always handle the constants separately from the variables. We already know that x/x=1, therefore:

$$\frac{2x}{x} = 2$$

Below is another example:

$$\frac{4x}{x} = ?$$

This problem is the same as the previous, therefore:

$$\frac{4x}{x} = 4$$

Below is another example:

$$\frac{4x}{2x} = ?$$

Here, we have a constant in the numerator and the denominator. We know that x/x=1, we also know that 4/2=2. Therefore:

$$\frac{4x}{2x} = 2$$

Below is another example:

$$\frac{6x}{2x} = ?$$

This problem is the same as the previous, except now we have 6/2=3 Therefore:

$$\frac{6x}{2x} = 3$$

Below is another example:

$$\frac{6xy}{2xy} = ?$$

Here, we already know that x/x=1, we also know that y/y=1. And of course we also know that 6/2=3. Therefore:

$$\frac{6xy}{2xy} = 3$$

Below is another example:

$$\frac{6x2y}{2x3y} = ?$$

Here, we have now introduced some extra constants. Therefore, we have to combine like terms before we can solve:

$$\frac{6 \cdot 2xy}{2 \cdot 3xy} = \frac{12xy}{6xy}$$

We know that xy/xy=1, and we know that 12/6=2, therefore:

$$\frac{6 \cdot 2xy}{2 \cdot 3xy} = \frac{12xy}{6xy} = 2$$

Below is another example:

$$\frac{6x3y}{2x3y} = ?$$

First we combine like terms, and then we solve:

$$\frac{6x3y}{2x3y} = \frac{6 \cdot 3xy}{2 \cdot 3xy} = \frac{18xy}{6xy} = 3$$

Below is another example:

$$\frac{xyz}{xy} = ?$$

Here, we have introduced a third variable z. We know that xy/xy=1. We also know that z/1=z. Therefore:

$$\frac{xyz}{xy} = z$$

Below is another example:

$$\frac{xy}{x} = ?$$

We know that x/x=1. We also know that y/1=y. Therefore:

$$\frac{xy}{x} = y$$

Below is another example:

$$\frac{6x3yz}{2x3y} = ?$$

First we must combine like terms:

$$\frac{6x3yz}{2x3y} = \frac{6 \cdot 3xyz}{2 \cdot 3xy}$$

Answer:

$$\frac{6 \cdot 3xyz}{2 \cdot 3xy} = \frac{18xyz}{6xy} = 3z$$

Below is another example:

$$\frac{3x4y2z}{4x2y} = ?$$

First we must combine like terms:

$$\frac{3x4y2z}{4x2y} = \frac{3 \cdot 4 \cdot 2xyz}{4 \cdot 2xy}$$

Answer:

$$\frac{3 \cdot 4 \cdot 2xyz}{4 \cdot 2xy} = \frac{24xyz}{8xy} = 3z$$

Below is another example:

$$\frac{x^3}{x} = ?$$

Whenever we divide, we subtract the exponents. Therefore:

$$\frac{x^3}{x} = x^{3-1} = x^2$$

Below is another example:

$$\frac{x^4}{x} = ?$$

Whenever we divide, we subtract the exponents. Therefore:

$$\frac{x^4}{x} = x^{4-1} = x^3$$

Below is another example:

$$\frac{x^5}{x^2} = ?$$

Answer:

$$\frac{x^5}{x^2} = x^{5-2} = x^3$$

Below is another example:

$$\frac{x^5 y^4}{x^2 y^3} = ?$$

Answer (remember y^1 is the same thing as y, y^1=y):

$$\frac{x^5 y^4}{x^2 y^3} = x^{5-2} y^{4-3} = x^3 y$$

Below is another example:

$$\frac{x^7 y^8 z^5}{x^2 y^3 z^2} = ?$$

Answer:

$$\frac{x^7 y^8 z^5}{x^2 y^3 z^2} = x^{7-2} y^{8-3} z^{5-2} = x^5 y^5 z^3$$

Below is another example:

$$\frac{x^5}{x^7} = ?$$

Here, we solve the problem the same as all the other problems. The end result is a variable raised to a negative exponent:

$$\frac{x^5}{x^7} = x^{5-7} = x^{-2}$$

Below is another example:

$$\frac{x^8 y^6 z^5}{x^7 y^4 z^7} = ?$$

Answer:

$$\frac{x^8 y^6 z^5}{x^7 y^4 z^7} = x^{8-7} y^{6-4} z^{5-7} = xy^2 z^{-2}$$

Below is another example:

$$\frac{2x^5 5y^7 6z^7}{3x^8 2y^3 5z^6} = ?$$

Answer:

$$\frac{2x^5 5y^7 6z^7}{3x^8 2y^3 5z^6} = \frac{2 \cdot 5 \cdot 6x^5 y^7 z^7}{3 \cdot 2 \cdot 5x^8 y^3 z^6} = \frac{60x^5 y^7 z^7}{30x^8 y^3 z^6}$$

$$= 2x^{5-8}y^{7-3}z^{7-6} = 2x^{-3}y^4z$$

Below is another example (remember $x^0=1$):

$$\frac{4x^8 5y^7 3z^7}{5x^5 3y^7 z^6} = ?$$

Answer:

$$\frac{4x^8 5y^7 3z^7}{5x^5 3y^7 z^6} = \frac{4 \cdot 5 \cdot 3x^8 y^7 z^7}{5 \cdot 3x^5 y^7 z^6} = \frac{60x^8 y^7 z^7}{15x^5 y^7 z^6}$$

$$= 4x^{8-5}y^{7-7}z^{7-6} = 4x^3 z$$

Below is another example:

$$\frac{x^5 3y^7 2z^2}{4x^4 3y^8 z^8} = ?$$

Answer:

$$\frac{x^5 3y^7 2z^2}{4x^4 3y^8 z^8} = \frac{3 \cdot 2x^5 y^7 z^2}{4 \cdot 3x^4 y^8 z^8} = \frac{6x^5 y^7 z^2}{12x^4 y^8 z^8}$$

$$= \frac{1}{2}x^{5-4}y^{7-8}z^{2-8} = \frac{1}{2}xy^{-1}z^{-6}$$

Solve the below problem:

$$x(x + 1) = ?$$

When multiplying with two terms inside the parenthesis, we multiply the term outside the parenthesis by each term inside the parenthesis. Since x multiplied by x = x², and 1 multiplied by x = x, we have:

$$x(x + 1) = x^2 + x$$

Below is another example:

$$x(2x + 3) = ?$$

Since x multiplied by 2x = 2x² and x multiplied by 3 = 3x, we have:

$$x(2x + 3) = 2x^2 + 3x$$

Below is another example:

$$x(2x^2 + 3x) = ?$$

Since x multiplied by 2x² = 2x³ and x multiplied by 3x = 3x², we have:

$$x(2x^2 + 3x) = 2x^3 + 3x^2$$

Solve the below problem:

$$(x + 1)(x + 2)$$

Here, we are multiplying two parenthesis together. Since there are four terms in total, we need to multiply four times. The multiplication is done in this order:

1. First terms first:

$$(x + 1)(x + 2)$$

$$x \times x = x^2$$

2. Outside terms second (x times 2 = 2x):

$$(x + 1)(x + 2)$$

$$x \times 2 = 2x$$

3. Inside terms third (1 times x = x):

$$(x + 1)(x + 2)$$

$$1 \times x = x$$

4. Last terms last (1 times 2 = 2):

$$(x + 1)(x + 2)$$

$$1 \times 2 = 2$$

Now we take our four answers and add them together:

$$x^2 + 2x + x + 2$$

Since we know 2x + x = 3x, we have:

$$x^2 + 3x + 2$$

Below is another example:

$$(x + 2)(x + 3)$$

First:

$$x \times x = x^2$$

Outside:

$$x \times 3 = 3x$$

Inside:

$$2 \times x = 2x$$

Last:

$$2 \times 3 = 6$$

Therefore we have:

$$x^2 + 3x + 2x + 6$$

Combining 3x and 2x gives us:

$$x^2 + 5x + 6$$

Below is another example:

$$(x + 4)(x + 6)$$

First:

$$x \times x = x^2$$

Outside:

$$x \times 6 = 6x$$

Inside:

$$4 \times x = 4x$$

Last:

$$4 \times 6 = 24$$

Therefore we have:

$$x^2 + 6x + 4x + 24$$

Combining 6x and 4x gives us:

$$x^2 + 10x + 24$$

Below is another example:

$$(2x + 7)(x + 5)$$

First:

$$2x \times x = 2x^2$$

Outside:

$$2x \times 5 = 10x$$

Inside:

$$7 \times x = 7x$$

Last:

$$7 \times 5 = 35$$

Therefore we have:

$$2x^2 + 10x + 7x + 35$$

Combining 10x and 7x gives us:

$$2x^2 + 17x + 35$$

Below is another example:

$$(3x + 5)(2x + 6)$$

First:

$$3x \times 2x = 6x^2$$

Outside:

$$3x \times 6 = 18x$$

Inside:

$$5 \times 2x = 10x$$

Last:

$$5 \times 6 = 30$$

Therefore we have:

$$6x^2 + 18x + 10x + 30$$

Combining 18x and 10x gives us:

$$6x^2 + 28x + 30$$

Here, every term is a factor of 2. While the above answer is correct, it is a common practice to reduce any term if possible. Therefore we divide every term by two, which gives us our final answer:

$$3x^2 + 14x + 15$$

Practice Problems

1. Combine like terms in the equation below:

$$6y^2 - 8x^2 + 2x^2 + 8 - 9y^2 + 3$$

Answer:

$$-8x^2 + 2x^2 = -6x^2$$

$$6y^2 - 9y^2 = -3y^2$$

$$8 + 3 = 11$$

Therefore:

$$6y^2 - 8x^2 + 2x^2 + 8 - 9y^2 + 3$$

$$= -6x^2 - 3y^2 + 11$$

2. Combine like terms in the equation below:

$$7x^2 - 7y^2 + 3x^2 + 7y^2 + 12$$

Answer:

$$7x^2 + 3x^2 = 10x^2$$

$$-7y^2 + 7y^2 = 0$$

Therefore:

$$7x^2 - 7y^2 + 3x^2 + 7y^2 + 12$$

$$= 10x^2 + 12$$

3. Combine like terms in the equation below:

$$x^2 - x^2 + x^2 + 4y^2 - 1$$

Answer:

$$x^2 - x^2 + x^2 = x^2$$

Therefore:

$$x^2 - x^2 + x^2 + 4y^2 - 1$$

$$= x^2 + 4y^2 - 1$$

4. Solve the below equation, with x=1 and y=7:

$$9y + 2x + 3x + y$$

First we combine like terms:

$$2x + 3x = 5x$$

$$9y + y = 10y$$

Therefore:

$$9y + 2x + 3x + y = 5x + 10y$$

Substitute x=1 and y=7:

$$5(1) + 10(7)$$

$5 + 70$

75

5. Solve the below equation, with x=2 and y=8:

$$-3x + 12y + 17x + 15y - 7$$

First we combine like terms:

$$-3x + 17x = 14x$$

$$12y + 15y = 27y$$

Therefore:

$$-3x + 12y + 17x + 15y - 7 = 14x + 27y - 7$$

Substitute x=2 and y=8:

$$14(2) + 27(8) - 7$$

$$28 + 216 - 7$$

$$237$$

6. Solve the below equation, with x=2, y=4 and z=3:

$$6y + 19x + 6z - 9x + 4y + 2z$$

First we combine like terms:

$19x - 9x = 10x$

$6y + 4y = 10y$

$6z + 2z = 8z$

Therefore:

$6y + 19x + 6z - 9x + 4y + 2z$

$= 10x + 10y + 8z$

Substitute x=2, y=4 and z=3:

$10(2) + 10(4) + 8(3)$

$20 + 40 + 24$

84

7. Simplify the equation below:

$5x^2 \cdot 6y^4 = ?$

Answer:

$5 \cdot 6 \cdot x^2 \cdot y^4 = 30x^2y^4$

8. Simplify the equation below:

$6x^{11} \cdot 7y^4 = ?$

Answer:

$$6 \cdot 7 \cdot x^{11} \cdot y^4 = 42x^{11}y^4$$

9. Simplify the equation below:

$$x^3 \cdot 3y^3 \cdot z^2 \cdot 4z^4 \cdot 2x = ?$$

Answer:

$$3 \cdot 4 \cdot 2 \cdot x^{3+1} \cdot y^3 \cdot z^{2+4} = 24x^4y^3z^6$$

10. Simplify the equation below:

$$\frac{x^5 9y^2 4z^3}{x^4 4y^2 z^7} = ?$$

Answer:

$$\frac{x^5 9y^2 4z^3}{x^4 4y^2 z^7} = \frac{9 \cdot 4x^5 y^2 z^3}{4x^4 y^2 z^7} = \frac{36x^5 y^7 z^3}{4x^4 y^8 z^7}$$

$$= 9x^{5-4}y^{2-2}z^{3-7} = 9xz^{-4}$$

11. Simplify the equation below:

$$\frac{x^2 2y^7 2z^2}{2x^3 4y^2} = ?$$

Answer:

$$\frac{x^2 2y^7 2z^2}{2x^3 4y^2} = \frac{2 \cdot 2x^2 y^7 z^2}{2 \cdot 4x^3 y^2} = \frac{4x^2 y^7 z^2}{8x^3 y^2}$$

$$= \frac{1}{2}x^{2-3}y^{7-2}z^2 = \frac{1}{2}x^{-1}y^5 z^2$$

12. Simplify the equation below:

$$\frac{6x^7 2y^9 7z^4}{4x^7 3y^3 6z^4} = ?$$

Answer:

$$\frac{6x^7 2y^9 7z^4}{4x^7 3y^3 6z^4} = \frac{6 \cdot 2 \cdot 7x^7 y^9 z^4}{4 \cdot 3 \cdot 6x^7 y^3 z^4} = \frac{84x^7 y^9 z^4}{72x^7 y^3 z^4}$$

If we divide both sides of 84/72 by 4, we get 21/18. And if we divide both sides of 21/18 by 3, we get 7/6:

$$= \frac{21}{18}x^{7-7}y^{9-3}z^{4-4} = \frac{7}{6}y^6$$

13. Simplify the equation below:

$$2x(3x^2 + 3x) = ?$$

Since 2x multiplied by 3x² = 6x³ and 2x multiplied by 3x = 6x², we have:

$$2x(3x^2 + 3x) = 6x^3 + 6x^2$$

14. Simplify the equation below:

$$4x(x^2 + 3x + 5) = ?$$

Since 4x multiplied by x² = 4x³, 4x multiplied by 3x = 12x² and 4x multiplied by 5 = 20x, we have:

$$4x(x^2 + 3x + 5) = 4x^3 + 12x^2 + 20x$$

15. Simplify the equation below:

$$(x + 2)(x + 2)$$

First:

$$x \times x = x^2$$

Outside:

$$x \times 2 = 2x$$

Inside:

$$2 \times x = 2x$$

Last:

$$2 \times 2 = 4$$

Therefore we have:

$$x^2 + 2x + 2x + 4$$

Combining 2x and 2x gives us:

$$x^2 + 4x + 4$$

16. Simplify the equation below:

$$(2x + 4)(3x + 2)$$

First:

$$2x \times 3x = 6x^2$$

Outside:

$$2x \times 2 = 4x$$

Inside:

$$4 \times 3x = 12x$$

Last:

$$4 \times 2 = 8$$
Therefore we have:

$$6x^2 + 4x + 12x + 8$$

Combining 4x and 12x gives us:

$$6x^2 + 16x + 8$$

Since we have only even numbers, we can divide by 2 to reduce to lower terms:

$$3x^2 + 8x + 4$$

17. Simplify the equation below:

$$(x + 2)(x + 3)$$

First:

$$x \times x = x^2$$

Outside:

$$x \times 3 = 3x$$

Inside:

$$2 \times x = 2x$$

Last:

$$2 \times 3 = 6$$

Therefore we have:

$$x^2 + 3x + 2x + 6$$

Combining 3x and 2x gives us:

$$x^2 + 5x + 6$$

18. Simplify the equation below:

$$(3x + 3)(4x + 7)$$

First:

$$3x \times 4x = 12x^2$$

Outside:

$$3x \times 7 = 21x$$

Inside:

$$3 \times 4x = 12x$$

Last:

$$3 \times 7 = 21$$

Therefore we have:

$$12x^2 + 21x + 12x + 21$$

Combining 21x and 12x gives us:

$$12x^2 + 33x + 21$$

Chapter 6 Solving Algebraic Expressions

When solving algebraic expressions, we are trying to find the variables in each equation that will make the expression be true. In pre-algebra, we only have to find one variable in each equation. Once we move to algebra, we will have to find multiple variables.

Below is the first example. For what value of x is the below equation true?

$$x + 2 = 3$$

The first step is to combine like terms. Here, we want all of the "non x" terms to be on one side of the equation, and we want all "x terms" to be on the other side of the equation.

Therefore, we have to move the 2 across the equation. Whenever we move a term across the equation, we have to change it from positive to negative, or vice versa. Here, we have to change "2" from positive to negative:

$$x = 3 - 2$$

$$x = 1$$

Below is another example:

$$x + 3 = 2$$

Here, we combine like terms and isolate x (move the 3 to the other side of the equation and make it negative):

$$x = 2 - 3$$

$$x = -1$$

Below is another example:

$$2x + 3 = x + 4$$

Here, we still combine like terms, but now we also move x across the equation and make it negative:

$$2x - x = 4 - 3$$

$$x = 1$$

Below is another example:

$$3x + 1 = 2x + 6$$

Same as before, we must combine like terms and isolate x:

$$3x - 2x = 6 - 1$$

$$x = 5$$

Below is another example:

$$2x + 3 = x + 2$$

Same as before, we must combine like terms and isolate x:

$$2x - x = 2 - 3$$

$$x = -1$$

Below is another example:

$$4x + 2 = 3x + 3$$

Combine like terms and isolate x:

$$4x - 3x = 3 - 2$$

$$x = 1$$

We can always perform an operation to the equation, as long as we perform the same operation to the other side of the equation. For example, say we want to multiply the above equation by 2. We can do this, but we have to do it to both sides of the equation:

$$2 \cdot x = 1 \cdot 2$$

This would give us:

$$2x = 2$$

2x=2 is the same as x=1. In both cases the solution to the equation is that x must equal 1.

We can also divide both sides of an equation. Here, we take the above equation and divide by 2:

$$\frac{2x}{2} = \frac{2}{2}$$

Here, we know that 2x/2=x, and 2/2=1. Therefore:

$$x = 1$$

When we divide one side of an equation by a number, we must also divide the other side of the equation by the same number. In this way, the equation is still equally balanced. Below is another example:

$$3x = 9$$

Here, we can divide both sides of the equation by 3:

$$\frac{3x}{3} = \frac{9}{3}$$

We know that 3x/3=x, and 9/3=3. Therefore:

$$x = 3$$

Below is another example:

$$3x + 2 = 4x + 3$$

Combine like terms and isolate x:

$$3x - 4x = 3 - 2$$

$$-x = 1$$

Here, we need to remove the minus sign from x. To do this, we can divide both sides by -1:

$$\frac{-x}{-1} = \frac{1}{-1}$$

Here, we know that -x/-1=x, and 1/-1=-1. Therefore:

$$x = -1$$

Below is another example:

$$4x + 3 = 2x + 7$$

Combine like terms:

$$4x - 2x = 7 - 3$$

$$2x = 4$$

Divide both sides to isolate x:

$$\frac{2x}{2} = \frac{4}{2}$$

Therefore:

$$x = 2$$

Below is another example:

$$5x + 5 = x + 17$$

Combine like terms:

$$5x - x = 17 - 5$$

$$4x = 12$$

Divide both sides to isolate x:

$$\frac{4x}{4} = \frac{12}{4}$$

Therefore:

$$x = 3$$

Below is another example:

$$7x + 13 =- 2x - 14$$

Combine like terms (when moving a negative number across the equation, we change the sign from negative to positive):

$$7x + 2x =- 13 - 14$$

$$9x =- 27$$

Divide both sides to isolate x:

$$\frac{9x}{9} = \frac{-27}{9}$$

Therefore:

$$x = -3$$

We have discussed combining like terms and dividing both sides of the equation to isolate x. We can also multiply both sides of the equation to isolate x:

$$\frac{x}{2} = 1$$

Here, we need to isolate x by removing the 2 from under x. To do this we can multiply both sides of the equation by 2:

$$2 \times \frac{x}{2} = 1 \times 2$$

We know that 2 * x/2=x, and we know that 1*2=2, therefore:

$$x = 2$$

Below is another example:

$$\frac{x}{3} = 2$$

Here, we can multiply both sides of the equation by 3:

$$3 \times \frac{x}{3} = 2 \times 3$$

We know that 3 * x/3=x, and we know that 2*3=6, therefore:

$$x = 6$$

Below is another example:

$$\frac{x}{6} = 3$$

Here, we can multiply both sides of the equation by 6:

$$6 \times \frac{x}{6} = 3 \times 6$$

*We know that 6 * x/6=x, and we know that 3*6=18, therefore:*

$$x = 18$$

Practice Problems

1. Solve for x:

$$4x + 3 = 3x + 8$$

Answer:

$$4x - 3x = 8 - 3$$

$$x = 5$$

2. Solve for x:

$$5x + 3 = 3x + 9$$

Answer:

$$5x - 3x = 9 - 3$$

$$2x = 6$$

Now divide both sides by 2 to isolate x:

$$\frac{2x}{2} = \frac{6}{2}$$

$$x = 3$$

. Solve for x:

$$5x - 6 = -4x + 9$$

Answer:

$$5x + 4x = 9 + 6$$

$$9x = 15$$

Now divide both sides by 9 to isolate x:

$$\frac{9x}{9} = \frac{15}{9}$$

$$x = \frac{15}{9}$$

Here, we can divide both 15 and 9 by three, therefore:

$$x = \frac{15}{9} = \frac{5}{3}$$

4. Solve for x:

$$5x - 15 = 7x - 7$$

Answer:

$$5x - 7x = -7 + 15$$

$$-2x = 8$$

Now divide both sides by -2 to isolate x:

$$\frac{-2x}{-2} = \frac{8}{-2}$$

$$x = -4$$

5. Solve for x:

$$\frac{x}{6} = 3$$

Answer:

$$6 \times \frac{x}{6} = 3 \times 6$$

$$x = 18$$

6. Solve for x:

$$2 + \frac{x}{10} = 7$$

Answer:

$$\frac{x}{10} = 7 - 2$$

$$\frac{x}{10} = 5$$

$$10 \times \frac{x}{10} = 5 \times 10$$

$$x = 50$$

7. Solve for x:

$$3 + \frac{x}{4} = 11$$

Answer:

$$\frac{x}{4} = 11 - 3$$

$$\frac{x}{4} = 8$$

$$4 \times \frac{x}{4} = 8 \times 4$$

$$x = 32$$

Questions or Comments?

his concludes Concise Pre-Algebra. We hope that you enjoyed this book nd found the content easy to understand.

 you enjoyed this book, please leave us a review on Amazon.

 you have questions or comments, email to 1fo@concisetextbooks.com.

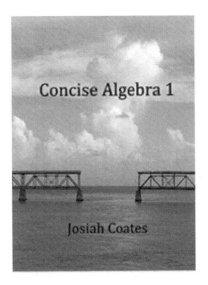

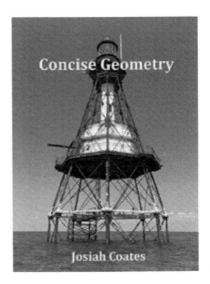

Made in the USA
Columbia, SC
29 January 2020